Mallory Myths
The Mallory Clothi[ng]

By Mike Parsons a[nd]

This brochure is published by the Mountain Heritage Trust

© Mountain Heritage Trust

ISBN 0-9553580-0-0   978-0-9553580-0-5

Mountain Heritage Trust
Station House
Ullswater Road
Penrith
Cumbria
CA11 7JQ

Cover: Everest (William Livingston)

Printed in Great Britain by
Blackmore Limited, Shaftesbury, Dorset

Contents

Introduction 4

1 Mallory Myths and Mysteries 5
2 The Reality behind the Myths and Photos 11
3 Project Inspiration 17
4 Rags to Replicas 18
5 Unveiling of the Mallory Replicas, September 2005 25
6 Everest Testing 29
7 John Hunt's Pyramid of Knowledge 32
8 Some Mallory Mysteries Resolved 36

Conclusion 42

Appendix 1 – Artifacts 43
Appendix 2 – Timeline 45
Further Reading 46
Acknowledgements 47
About the Authors 48

Introduction

THIS booklet tells the exciting story of the replication of clothing worn on the 1924 Mount Everest expedition, its testing on Everest in 2006 and the new light this throws upon the mystery of George Mallory and Sandy Irvine, who perished close to the summit in June 1924. The story unfolds against the backdrop of emerging mountaineering and equipment innovation and shows these climbers to be amongst the pioneers of high altitude, lightweight climbing. It is up to readers to decide whether or not Mallory and Irvine reached the summit of Everest, but we invite them to think again about the conventional view that their clothing was an impediment to their success.

1 Mallory Myths and Mysteries

On 8th June, 1924, Britain's finest and most expert mountaineer, George Leigh Mallory and his young climbing partner, Sandy Irvine, met their untimely deaths attempting to reach the summit of the world's highest mountain, Mount Everest. The first undisputed ascent was not until 1953, by Edmund Hillary and Tenzing Norgay, members of the British team led by John Hunt. However, the possibility that Mallory and Irvine actually reached that elusive summit in 1924 before meeting with disaster continues to intrigue climbers and researchers. As Everest historian, Audrey Salkeld says, 'George Mallory remains an enigma, despite the discovery of his body on 1 May 1999. Mallory, with his public school education and First World War experience, epitomises modern images of inter-war climbers and climbing.' Fascination with the 1920s Everest climbers continues, but often these men are depicted as ill-prepared and under-equipped in their search for their 'wildest dream'.

The first Everest expedition with Mallory, back row, far right. The myths begin here, but clothing evolution is also evident: note the thigh-length flying boots. (Royal Geographical Society)

Myths reinforced. But look more closely at Sandy Irvine, far left, back row. His jacket is not tweed, but Burberry windproof and he has sewn recent innovations – zips – to his pockets. Norton, centre, back row, is also wearing a tailored Burberry jacket. Howard Somervell, front row, third from left, has soft, fur-lined ethnic-style boots. (John Noel Collection)

The famous 1920s Everest base camp photographs of men in tweed Norfolk jackets, breeches and 'hobnailed' boots displayed at the *National Mountaineering Exhibition* attract comments like: 'Well, with clothing like that they didn't stand a chance on Everest, did they?'

Everyone wears clothes and so non-climbers have opinions on the Everest clothing because it is easier to understand and judge than technical issues such as oxygen and dehydration. In the 1920s people were sceptical about the climbers' chances of success and the author and playwright George Bernard Shaw described the 1922 Everest expedition as looking like a 'Connemara picnic surprised by a snowstorm'. Many with later Everest experience dismissed the 1920s kit as inadequate. On completing his solo ascent of Everest in 1980, Reinhold Messner said that he did not believe Mallory and Irvine could have reached the summit because of their poor clothing and equipment.

Interest in the Mallory and Irvine mystery culminated in a joint Anglo-American Research Expedition to the north side of Mount Everest in 1999. On 1 May Conrad Anker, a member of the expedition team, found George Mallory's body at 8155m, frozen into the rock and scree below the north ridge of Everest. Attracted by a patch of intense white on the mountainside, he climbed nearer and 'suddenly saw hobnailed boots and old clothing'. Clothing name tags, letters and monogrammed handkerchiefs revealed that he had found George Leigh Mallory himself.

The discovery brought a poignant reminder of the apparent flimsiness of the clothing worn on Mount Everest in the 1920s. Looking from his own modern perspective at Mallory's outfit of leather nailed boots, three pairs of hand knitted stockings, long underwear and puttees and seven or eight layers of silk, cotton or wool and fur-lined flying helmet, Anker was struck by the contrast with his own outfit which consisted of:

- two layers of fleece,
- a synthetic woollen parka,
- a full down suit with wind-resistant surface,
- a knitted hat under the built-in down hood on his jacket,
- thick nylon boots insulated with closed-cell foam, with gaiters built in to keep snow out of the ankles.

Anker's down suit alone provided three to four inches of insulation, which is more than all of Mallory's layers combined.

For Anker, and for many others, poor clothing was one of several odds stacked against Mallory and Irvine – along with the capricious oxygen sets, a lack of understanding of the hazards of dehydration and the technical difficulties of climbing the rocky barrier on the route to the summit known as the Second Step. A few people have recognised that the clothing used in 1924 was the best available at the time, but up to now it has been assumed that modern clothing is much better.

Artifacts recovered by 1999 Mallory and Irvine Expedition.
(Royal Geographical Society)

The Puzzles of the 1924 Everest Clothing and Footwear

There are many puzzles and questions surrounding the clothing and equipment used in 1924 that this brochure aims to unravel:

1. **Did they really use 'Norfolk' style tweed jackets as all the old photographs show?**
   Was everything quite as it seemed in those old photos? Are the photos really a window on 1920s climbing dress, or are they posed images, to be sent home for the record?

2. **Was it possible that they were warm enough without down clothing?**
   Down clothing (fluffy duck down sandwiched between 2 waterproof layers) is now assumed to be crucial for high altitude climbing. All the modern photographs show Himalayan climbers looking a little like Michelin men.

3. **Why did the climbers not use a hooded jacket as Roald Amundsen did in Antarctica 15 years earlier and others before him as early as 1875?**
   Are not hoods essential on high performance clothing for high altitude climbing?

4. **Were the Mallory layers very different from today, and if so, in what way?**

5. **Why didn't they wear crampons?**
   How could the climbers manage on snow and ice without crampons? Surely they would have found it difficult to keep their footing? Had crampons been invented by 1924?

6. **Surely those heavy nailed boots could not have been suitable for the ascent?**
   Did we not change to rubber-soled boots because they performed better?

7. **How did they keep their feet warm?**
   Modern boots are very well insulated, but surely those old leather boots could not have kept their feet warm? So did they get frostbite long before they reached the summit?

2 The Reality behind the Myths and Photos

The 1920s expedition photos are deceptive. In the main they were taken at base camp, not on the climb, and do not necessarily show what people wore when climbing. Also the 1920s was a less casual age and photographic equipment more cumbersome, so that the photographs were likely to be more formal and stilted than ours today. Like all explorers and pioneers, there was much the 1924 Everest mountaineers did not fully understand – for example, how to increase oxygen supply effectively and the problem of high altitude dehydration. But what they did understand, perhaps better than modern climbers, was how their clothing functioned, what was needed and what was possible.

Their knowledge came from a combination of published travel advice, pre-First World War polar and high altitude exploration, clothing suppliers and the earlier 1920s Everest expeditions. For a start, the

Mallory, left, and Irvine. Last picture of them alive as they prepare to leave Camp 6.
(Royal Geographical Society)

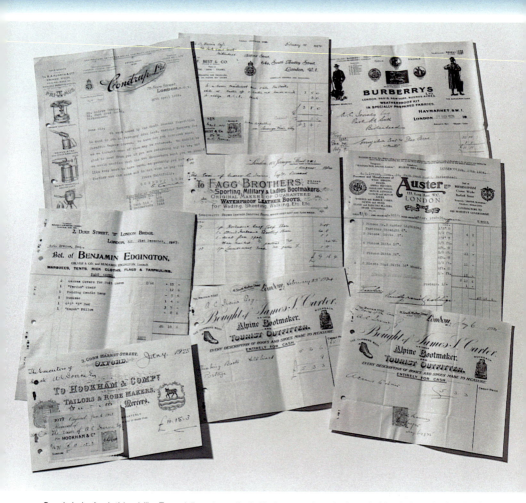

Sandy Irvine's clothing bills. Especially noteworthy is Burberry – where he bought his windproofs, Condrup – the UK agents for Primus, and Auster – where he ordered his 'lightning' zips. Benjamin Edgington, founded in the eighteenth century and makers of many famous tents, including the Whymper and the Mummery, became part of Blacks. (Sandy Irvine Trust)

climbers had more than just experience in the Alps to draw upon, as there was the nineteenth century knowledge of climbers such as Edward Whymper and Frederick Mummery who had climbed in the greater ranges. There was more than second-hand knowledge available, of course, although George Mallory and several others on the early Mount Everest expeditions had not themselves climbed outside the Alps. Equipment officer on the 1922 expedition, Charles Meade had reached over 7000 metres in the Himalaya before the First World War, as had several of the 1920s Everest expedition members. Charles Bruce was a lightweight specialist, while Tom Longstaff and Alexander Kellas had both made high altitude climbs.

Everest replaced the North and South Poles in the public's imagination, taking centre stage on the Royal Geographical Society (RGS) and Alpine Club's agendas after the First World War. The RGS stimulated exploration and from 1854 published *Hints for Travellers*. This covered polar travel and mountaineering, giving detailed lists of what to take and where to obtain it and was regularly updated. Technical climbing equipment was almost always sourced from abroad and so were rucksacks. But tents and clothing were made in Britain, often adapted to climbers' own specifications.

Modern climbers build a close relationship with their suppliers and it was no different in 1924. The expedition members approached specialist manufacturers, tailors and boot-makers such as Benjamin Edgington, James S. Carter and Silver and Co. Sporting Outfitters. The difference was that their suits were of a standardised design adjusted to the personal measurements of the climber. The 1924 climbing team were instructed, when they went to Messrs Burberry in Haymarket, to ask for Mr Pink for a careful fitting, crucial if the outer-garment was to fit over multiple layers. This was all part of the dialogue that connected the climbing community to their suppliers and to earlier polar expeditions. Himalayan climbing was in its infancy, but mountaineers understood that they would experience similar climatic conditions to the poles – dry, windy and cold – but with the additional consideration of altitude.

Knowledge certainly grew with each successive expedition and, by 1924, practical experience of the conditions, combined with improvements and adaptations to polar kit, demonstrated progress and understanding of the requirements for Himalayan climbing. The sketchy kit-list for 1921 is replaced in 1924 by a detailed, well-researched set of guidance notes, which drew heavily on the experience and research by climber and scientist, George Finch, from the 1922 expedition. Edward Norton's 1924 Everest list described a layered clothing system for active climbers at high altitudes and was adapted from the combination of polar wind-proofs, insulation and air-trapping layers.

Sandy Irvine, Spitzbergen, 1923. (Fell and Rock Climbing Club Library Collection)

George Finch, the 1922 expedition equipment expert, was quite clear that there were two clothing zones on Everest. Up to the North Col at 7010 metres, Alpine mountain clothing was adequate if the climber was moving. Above it the vicious winds ripped through everything, making high performance wind-resistant garments vital. He suggested that a thin rubber coating be applied to the windproofs, though this would have impeded breathability and was never used.

A little-known photograph of Sandy Irvine in Spitzbergen, Northern Norway, in 1923 offers a different image of cold weather clothing from those at Everest base camp. Apart from his flying helmet and pipe he looks remarkably modern and is clearly dressed in high performance clothing. He is wearing an adaptation of the Shackleton windproof sledging suit used in Antarctica and well-insulated footwear.

George Mallory Everest Base Camp 1924.
(John Noel Collection)

With a closer examination of the 1924 base camp team photograph some of the myths start to fade. Sandy Irvine (back row, far left) had sewn zips onto the pockets of his jacket. Zip fasteners were cutting-edge technology at the time. Also the jacket is made in a superior Burberry windproof fabric similar to that Shackleton had used in Antarctica and Irvine had worn the year before in the Arctic. Of the other team members, at least four were wearing Burberry jackets with Norton's being an individually tailored one, first developed by Burberry before the First World War.

3 Project Inspiration

The first sight of the fragments of Mallory's clothing brought back from Everest was deeply moving and inspired the authors to undertake a research project that would draw on their combined forty years of experience in manufacturing modern outdoor clothing plus their academic knowledge of this industry. However, they also recognised that scientific analysis would bring improved understanding of the performance of Mallory's clothing. The fragments were in too poor a condition to be displayed in a museum environment, but they held many of the secrets of the 1924 clothing.

The authors began by approaching the Textile Conservation Centre (TCC), which is part of the University of Southampton, for advice on preserving the textile fragments. They also enquired as to whether it would be possible to replicate them. Their preliminary report in October 2001 provided the foundation for a Heritage Lottery Fund bid prepared by the authors of this brochure on behalf of Mountain Heritage Trust (MHT), the current custodian of the clothing fragments. The bid was signed by John Innerdale and Sir Chris Bonington on behalf of the Trust in September 2002 and funding of £29,355 was approved in April 2003, leaving the balance of £3300 to be raised from other sources. This came from a Pasold Research Fund grant and surpluses from the Clothing for Extremes Conference organised by the authors of this brochure, who ran the project on a voluntary basis.

1924 Everest Expedition: Team member in the ice pinnacles of the East Rongbuk Glacier. (Royal Geographical Society)

4 Rags to Replicas

It took nearly three years of intensive scientific analysis (including scanning electron microscopy and microphotography, spectrometry and X-ray spectroscopy) and detective work to transform the fragments brought back from Everest into testable replicas. This work involved teams from the four universities of Southampton, Derby, Leeds and Lancaster.

How was it done?

The project was not just about a set of textile fragments, or even a set of replicated clothes, but about understanding their position in the history of mountaineering and the wider innovation process. This unusual and collaborative project involved historians of innovation, textile conservators and replica makers, working alongside performance clothing specialists, textile and clothing manufacturers and mountaineers. The replicated fabrics and garments all had to meet precise specifications and the decline of the Lancashire cotton industry since the 1920s made this particularly challenging. Several fabrics, such as the Burberry windproof, the wool flannel and the silk woollen vest, the cotton long-johns and the puttees, were custom made and many garments were hand-knitted. Modern silks had to be specially treated, or de-gummed, to match 1920s processes.

All contributors to the project were committed to conserving the textile fragments. So testing was non-intrusive and did no damage. The aim was to produce a testable replica that would perform just like the original clothing on Everest rather than simply garments for museum display.

Before full replication began, the experts were curious to understand how windproof the fabrics were and what kind of insulation they gave. So the first destination was the Performance Clothing Research Centre at University of Leeds where Dave Brook, an expert in analysing modern textile performance, tested the materials thoroughly exactly as he would modern fabric samples. First he measured the TOG value (the same measure used for assessing the warmth of a duvet) and found that together the layers were equivalent to 3.5 TOG, which was the insulation worn by Ranulph Fiennes and Mike Stroud when they were working hard at -40 degrees C in Antarctica in 1996. Brook also tested the jacket fragments and found they were very close to modern Pertex fabric in terms of windproof quality. He concluded that Mallory was sufficiently well insulated to survive on Everest, provided he was moving, though not for a bivouac. The first signs indicated that the fabrics were high performance and that the layering of the clothing was effective.

The primary fabric and fibre analysis was carried out by a team of researchers led by Amber Rowe at the Textile Conservation Centre (University of Southampton). The biggest puzzle was the vest, because it was shiny and looked as though it was made of mercerised cotton. This would be unsuitable for a base layer, because cotton is uncomfortable next to the skin since it absorbs perspiration and then chills the wearer rather than transporting the moisture away. However analysis revealed that the garment was a mixture of wool and silk. This combination of light and heavy denier fibres has become common in base layers only in the last 10 years and is usually known as 'denier gradient', making the 1924 vest especially impressive. It gives improved directional moisture transfer.

According to Amber Rowe of the Textile Conservation Centre, Mallory was wearing a hand-knitted jumper, two other hand-knitted fragments, one of which may be part of his leggings, three pairs of socks, a blue and white shirt, a vest that is in a fibre blend with silk, and his puttees which might be cashmere rather than sheep wool.

This was not based on casual visual observation because, as Rowe says, 'An expert can identify the breed of animal by the shape and patterning of the scales as well as other features'. There were also two silk shirt layers and a wool flannel shirt.

Analysis confirmed that Mallory was wearing one of the most high performance cotton fabrics of the age, made into a Burberry windproof climbing suit and worn as the outer layer. The first giveaway was a tiny fragment of label. The fabric was made of densely woven fine cotton.

The detective work

Vanessa Anderson at work on the replication process.

Rags to Replicas

21

Mary Rose, Mike Parsons and Joyce Meader deep in thought at the TCC laboratories.

Burberry has not always been the fashion brand we know today. Thomas Burberry, a Hampshire outfitter, had begun making sportswear for hunting and shooting in 1865. Described as 'self ventilating', the breathable fabrics used in Burberry clothing were the nineteenth century equivalent of Gore-Tex®.

Garment Design

Once the researchers knew what the garments had been made of, they needed to be sure what they looked like and the fragments only provided patchy evidence. Turning a fragment into a pattern and finally into a replica involves laying threads along the grain of the fabric to work out the exact dimensions and measurements. This is followed by careful study of construction features, such as seams and hems. Visits to archives and museums plus extensive photo-research were needed to see the kinds of clothing popular with mountaineers in the 1920s before a pattern could be drawn. The accuracy of the knitwear was much enhanced by replica maker Joyce Meader's extensive collection of historical knitting patterns. The replica makers did much more than just follow patterns – they helped to shape them. They worked closely with Amber Rowe at the TCC, with Dave Brook in Leeds, and with the authors, to unravel further secrets of the fragments. The finished garments are fully testable and show an extraordinary attention to detail of styling and finish, right down to buckles and buttons.

Some fragments were harder to identify than others, but the hardest was what became known as the 'brown knitted thingy'. This was a small piece of knitting with a fold along one side. Joyce Meader was finally able to identify it from some of the stitching detail as a *convertible muffler helmet* or *cap comforter* found in patterns used in the First World War. It is the predecessor of the modern Buff.

As already established, George Mallory was not dressed in tweeds when he died, but the style of his windproof jacket was unknown and a Norfolk jacket was initially assumed. By researching behind the fragments Vanessa Anderson, from the University of Derby, who was responsible for replicating the shirts and windproofs, found that he was, in fact, wearing a Burberry 'Everywhere' jacket as part of a climbing suit. This high performance sportswear included a pivot, or articulated, sleeve that had been patented by Burberry in 1901. This was crucial for free movement in sports, making it ideal for rock climbing and mountaineering.

Numerous fashion shifts make 1920s clothes appear dated, but body shape has changed too in the last 80 years. This affects the design, fit and performance of the garment as Vanessa Anderson explained after completing the shirts and outer-wear:

> The fit historically has always been a lot tighter ... and that generation were trained to stand very straight-backed and chest out from childhood, as seen in photos of the period. Our posture and sizing has changed. We slouch. If you try on garments from the Second World War and earlier the same size will feel a lot tighter.

This is confirmed by the slightness of Mallory's measurements and the tightness of his garments for an athletic man of the same height:

Height: 5' 11"
Weight: 11stone 5lbs
Chest: 35"–37"

This made replication more complicated and knitter Joyce Meader recalled:

> Adapting old patterns can often be difficult as the number of stitches required does not always match the shape of the 'vintage person'. This was proved to my dismay when the brown long-johns, on the first knitting, came up as a size 48/50" waist, and after spending one working month making a whole leg I had to unpick it and start again. So these became known as 'the three legged undies'.

5 Unveiling the Mallory Replicas September 2005

With the garments completed, it was obvious that this was a sophisticated and extremely light clothing system. But how did it compare with modern garments? This was the key question when the garments were unveiled by mountaineer Alan Hinkes at the Clothing for Extremes conference held at the Rheged Discovery Centre in September 2005. The story of the 1924 Mount Everest expedition remains one of the most enduring mysteries in climbing history, to which the Mallory clothing project, with its link between history and science, between clothing and mountaineering, gave a new and important twist. For the first time the clothing worn in 1924 was viewed, not as a relic of a bygone age, but as innovative high performance gear, which compared well with modern mountaineering clothing. There was, of course, a high level of national media interest in the unveiling.

1924 meets 2005. Tony Breakell and Alan Hinkes at the Mallory replica unveiling, Rheged Discovery Centre, 28 September 2005. (Gwen Ainsworth)

The Mallory Layering System

Clothing Weights
- **Upper Body**

Silk wool vest	140 g
Silk shirt (beige)	342 g
Shetland pullover	314 g
Silk shirt (green)	248 g
Flannel (wool) shirt	595 g
Burberry jacket	824 g
Sub total	**2675 g**

- **Lower body**

Cotton long-johns	275 g
Green Shetland long-johns	320 g
Brown Shetland long-johns	450 g
Burberry breeches	440 g
Sub Total	**1485 g**

Total ... **4160 g**

- **Footwear**

Blue socks	108 g
Mixed Shetland socks	82 g
Argyle socks	82 g
Puttees	106 g
Sub Total	**356 g**

- **TOTAL WEIGHT excluding boots** ... **4516 g**

The boots were not replicated but, at half the weight of modern boots, these were the lightest high altitude boots ever to be worn on Everest. The boot was based primarily on George Finch's design and made extensive use of polar knowledge. According to Sandy Irvine, some of the high altitude boots disintegrated, but Mallory's did not. The thick wool felt boot was overlaid with calfskin and had a 10mm felt mid-sole and a 3mm leather sole. Unlike conventional nailed boots, nails were pre-attached only as deep as the felt mid-sole to avoid heat being conducted from the feet by the metal. These nails were also lighter than conventional mountaineering boot nails. Heavy footwear is a major drain on energy, especially at altitude. Harold Raeburn, who had attempted Kangchenjunga in 1920 and was responsible for equipment in 1921, had carefully calculated:

> As regards weight, it is often not realised what a handicap heavy footgear is. A calculation shows that, given a difference of 2lbs between the boots of 2 climbers A and B, making an ascent of Ben Nevis (in Scotland) the heavier shod drags off the ground more than 9 tons than the lighter. The distance is 7 miles, and the height more than 4,400 feet. In the case of Mont Blanc from Chamonix the difference will come to nearly 26 tons.

The implication for climbing Everest can only be imagined. It was a lesson that the 1924 Everest climbers took to heart, as Mallory's boot shows.

High Altitude Boots on Everest 1924-2004

Boot description	Boot weight per half pair	Crampon weight	Total weight
1975, Galibier Makalu double boot, neoprene over-boot	1600 g	600 g	2400 g
Late 1970s Plastic shell boot with closed cell foam lining	1400 g	600 g	2000 g
2004 Scarpa 8000 with integral gaiter	1400 g	600 g	2000 g
1953 Everest boot	1200 g	600 g	1800 g
George Mallory boot 1924	800 g	No crampon	800 g

These findings raised the question of how modern gear would compare with the kit worn by George Mallory.

Typically a modern outfit today would weigh more than Mallory's and our careful questioning showed a typical weight of 4800/4900 grams for modern gear to summit 8000m peaks. This changed the way we perceived the clothing, the 1924 climbers' knowledge and their achievement. They were not merely using the most advanced garments available. They recognised that, especially at altitude, weight sapped energy and reduced the chances of success.

Project Team and special guests at the Mallory replica unveiling, Rheged Discovery Centre, 28 September 2005. Back row, left to right: *Dave Brook, John Angus, Mike Parsons, Terry Gifford, Jochen Hemmleb, Mary Rose, Joyce Meader, Amber Rowe, Vanessa Anderson.* Front row: *Tony Breakell in Mallory outfit and Alan Hinkes in 2005 outfit.* (Gwen Ainsworth)

6   Everest Testing

We could not know how well or badly these clothes actually felt compared with modern clothes until they had been field-tested. Graham Hoyland, BBC producer and great nephew of Howard Somervell, veteran of the 1922 and 1924 Mount Everest expeditions, became interested in the Mallory clothing project after hearing about the unveiling. Hoyland had climbed Everest himself in 1993 and provided the inspiration for the 1999 Mallory and Irvine Research Expedition, making him an ideal person to test the garments. He remembers meeting Somervell and listening to his vivid account of climbing high on Everest with George Mallory in 1922.

82 years after the ill-fated 1924 expedition, Graham Hoyland tested the clothing first at 3658 metres at Everest Base Camp and again at 4877 metres on the Rongbuk Glacier. His reports, received by e-mail on 27 April and 6 May 2006 from Everest Base Camp, were clear and revealing. On dressing in layers prior to testing, he was especially struck by the contrast between these natural-fibre garments and modern polypropylene. 'Warm to slip into rather than cold and clammy,' he wrote. His report confirmed that this was, indeed, an advanced clothing system:

> When exposed to a cutting wind blowing off the main Rongbuk glacier, I found the true value of the Burberry outer layer. This resisted the wind and allowed the eight layers beneath to trap warmed air between them and my skin. The patented pivot sleeve of the jacket enabled me to lift my arm right above my head (as when, say, striking a blow with an ice-axe) without displacing any layers. One immediate problem would be buttons and cold fingers: I suspect they would have put the clothes on at Advanced Base Camp and left them on for the duration. Fly buttons may have been left undone: there are enough layers to interleave. On the glacier the clothes felt extremely comfortable. When cutting steps with John Innerdale's long-handled ice-axe I found I could lift my arm to full extent without disturbing the

warm layers of air. I tested the whole suit with the expedition leader wearing his modern down suit next to me. Compared to the modern suit mine was far less cumbersome and I could squeeze through a tent flap easily. Both of us got too hot working on the glacier, so we both felt the Mallory clothing was more than adequate to climb to the summit in. We both thought it would be hard to survive a bivouac near the summit, though.

His conclusion that the clothes were very pleasant to wear, easy for movement and sufficiently warm to summit, finally lays to rest the myth that clothing was a major impediment to George Mallory and Sandy Irvine's attempt to climb Everest. As one recent mountain gear maker

Graham Hoyland, Rongbuk Glacier April 2006. He concluded that these comfortable garments were sufficient for a summit day. (Jen Peedom)

Pastel painting of Gauri Sankar by Howard Somervell, 1924 Expedition (Royal Geographical Society)

said, 'The best gear is invisible. You need to be able to forget about it and get on with the job'. The comfort which Graham describes suggests that the Mallory 1924 clothing more than met those criteria, although it was insufficient for an enforced night bivouac high on the mountain.

7 John Hunt's Pyramid of Knowledge

The 1924 expedition's experience with clothing, footwear and equipment contributed to success on Everest in 1953. In recognition of this, John Hunt paid especial tribute to George Mallory and George Finch, whose practical approach to Mount Everest he believed was so valuable to later expeditions. There was, he said, 'a pyramid of knowledge [and experience] from every attempt, each adding to the last until the puzzle is solved' and he saluted George Finch for his contributions, including footwear, oxygen and clothing. Innovation is evolutionary and, in fact, the pyramid, as it relates to lightweight clothing and equipment, can be traced back to the 1890s.

The project has shown that Mallory and Irvine and their 1924 companions were using a lightweight approach to clothing and equipment. It challenges the assumption that modern clothing and equipment is lighter and less cumbersome than that used in 1924. We might find it hard today to think of Mallory or any of his contemporaries as lightweight specialists, but that was precisely what they were. Norton's comments on the nailing of the boots were especially telling: 'Boots should be nailed sparingly for lightness, every ounce counts'. With weight-saving in mind, Sandy Irvine, who had taken on George Finch's role as expedition gear repairer and oxygen mechanic, made a very light rucksack using one of the oxygen frames for summit attempts.

Mallory himself summed up his thinking on weight when he wrote,

> It is unthinkable with this plan I shall not go to the top ... My intention is to carry as little as possible, move fast and catch the summit by surprise.

The clothing that George Mallory and Sandy Irvine wore was genuinely innovative and part of an evolutionary design process, reaching back into the nineteenth century. Lightweight materials, clothing and equipment are not just 21st century phenomena, but were developed for sport well before the First World War, using natural, rather than synthetic fibres.

Edmund Hillary putting on his high altitude boots, 1953. (Royal Geographical Society)

Early mountaineers, cyclists and explorers experimented with materials, developed patterns and worked with craftsmen on new designs, and this dramatically reduced the weight carried or worn.

Frederick Mummery, for example, was a nineteenth century lightweight pioneer and the oiled silk version of his classic tent weighed less than 2lbs. The next generation of lightweight pioneers included Alpinists W.T. Kirkpatrick and Philip Hope. These enthusiastic guideless climbers chose clothes that combined lightness and utility, using Shetland wool and silk – just as George Mallory was wearing 20 years later – and Hope developed a rucksack weighing only 7oz. Their lively articles in the *Alpine Journal* before the First World War will surely have inspired others. Tom Longstaff, who was a member of the 1922 Mount Everest Expedition, was one of the most imaginative of the early lightweight specialists and his experience will have been invaluable for the 1924 Everest expedition. He pioneered the use of silk rope, which he described as being very light, but very expensive, and used lightweight equipment on Nanda Devi in the Himalaya in 1903.

If these earlier enthusiasts of the lightweight approach provided the examples which Mallory's clothing and footwear built upon, later climbers, in their turn, developed Mallory's philosophy and practice further, right up to John Hunt's successful 1953 expedition. Although the boot in use in 1953 was heavier than Mallory's, Hunt's faith in lightweight footwear was endorsed by his expedition physiologist Griffith Pugh's calculation that was an echo of Raeburn 29 years earlier: 'in terms of physical effort, one pound's weight on the feet was equivalent to five pounds on the shoulders'. Finch's ideas for Everest boots were used in 1924 to combat frostbite and elements of his design appeared again in both the general climbing boots and the high altitude boot in 1953. Robert Lawrie had supplied the 1930s Everest expeditions and designed a general climbing boot weighing 1700 g 'lined with opossum fur between two layers of leather with a woollen felt sole' – as used in 1924 – but with a thin rubber sole. Other elements of Finch's design – particularly the outer protective layer – look remarkably like a predecessor to the high altitude SATRA boot used in 1953. As Charles Wylie of the 1953 expedition observed: 'We enjoyed the advantage of light boots throughout the expedition and there were no cases of frozen feet.'

Faith in lightweight clothing continued through to 1953 as more choices of material became available. There had, for instance, been continued

Kendal Mint Cake: Polar and Everest Expeditions were intertwined. (Jochen Hemmleb)

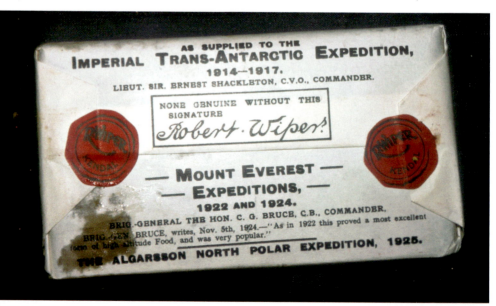

fabric and garment design developments with hooded, zipped Grenfell suits replacing Burberry for Everest in the 1930s. Highly functional, sophisticated down clothing was developed first in Continental Europe in the 1930s by climber Pierre Allain, where it was vital for the technical big wall climbing which was pioneered there, when bivouacs became routine. This clothing was further improved for mountain warfare and available from Continental suppliers by 1953 when it was combined with cotton-nylon windproof suits. Although the materials had changed, the principles of layering – the different functions of each complementary layer – had been retained from Mallory's time.

8 Some Mallory Mysteries Resolved

B eginning with the tweed Norfolk jacket, one by one the myths about 1924 Everest clothing have fallen away and just a few puzzles remain.

'A silk rope is useful in many ways to the guideless climber. For instance it will obtain civility from haughty hotel servants and even guides themselves may be beguiled into conversation by its means.'
Tom Longstaff, camp on Nanda Devi, 1903.
(Fell and Rock Climbing Club Library Collection)

1. Did it matter that they did not have down suits?

Mallory and Irvine were not under-equipped for a rapid summit attempt. Going lightweight means carrying slightly less than you are mentally comfortable with, and that is precisely what they did. So there was absolutely no margin for error – no chance of surviving a forced bivouac – because there was no down clothing for protection and warmth. Down provides insulation and is a psychological comfort in freezing conditions. It is a marvellous insulator and vital for a bivouac at altitude. But, as Graham Hoyland says, for active climbing it can be too warm, causing sweating and then chilling.

George Finch had experimented with down on the 1922 Everest expedition. His eiderdown lined coat, trousers and gauntlets were covered in balloon fabric and, whilst his companions thought it a huge joke, he was impressed by its performance. On 12 April 1922, during the march in, he wrote in his diary:

> The wind was poisonous from the beginning. I wore flying boots ... we also wore all manner of warm head-gear. Everyone felt the cold except myself – my eiderdown coat, trousers, flying boots and flying helmet kept me as warm as toast all through... [while six days later he confirmed] Everybody now envying me my eiderdown coat and it is no longer laughed at. May it do its job well on Everest.

2. So why was down not used in 1924?

Finch's eiderdown suit was listed in Norton's 1924 equipment notes, but no one took it. Even Sandy Irvine, with his enthusiasm for anything new, had struck it off his kit list. No one knows exactly why it was abandoned, but part of the explanation may have been suspicion of its champion. Finch, with his unconventional approach and Australian background, was dropped from the team because his face didn't fit with the elderly grandees of the RGS and the Alpine Club. There may have been other reasons, because it is not clear how robust the suit was for hard climbing. Down's absence in 1924 would not have stopped Mallory and Irvine reaching the summit. But after their accident they had no protection from the cold.

3. Why didn't they have hoods?

We take hoods for granted as an integral part of modern outdoor jackets. Hoodless jackets mean greater exposure to wind and snow in the higher chest and neck and so their absence in 1924 was a puzzle.

This weak point is probably exemplified by the fact that Somervell almost died of a frozen larynx on his summit attempt in 1924. Mallory was wearing a fur-lined flying helmet and a hoodless jacket. Hooded garments – or anoraks – are normal wear for the indigenous peoples of the arctic. Hoods were a comparative rarity outside Polar regions and George Finch has been credited as the pioneer of the hooded anorak for Alpine use in 1909. Burberry had produced special detachable polar helmets before the First World War, but climbers need a greater degree of peripheral vision than is required for polar travel. The porters' suits were hooded and a few of those with Polar experience had hooded jackets, but it is likely that most climbers did not like them, preferring the good visibility provided by the flying helmets. The leather flying helmets developed for the Flying Corps during the First World War and used in 1924 were fur lined and would have been very warm, indeed too warm at times because they were using lighter hats in several photos. UV-resistant sun creams did not exist. Helmets would possibly have given better protection than the hood designs of the time, and complete face-masks were often used, although rarely photographed. Irvine suffered severely from sunburn, losing a lot of skin from his face, which made wearing his oxygen mask painful.

4. Were the Mallory layers very different from today and, if so, in what way?

The concept of multiple layers, rather than a few thicker layers, came from the polar explorers. It was transferred to the Alps and the Himalaya and is widely accepted today. What was different about the layering then, compared with today was that when silk (windproof) was layered over wool (insulation), not only was the insulation enhanced, but the wind resistance also. Even before all the garments were replicated the experts had a sense that layers would slide on each other and create a sense of freedom of movement. This is exactly what Graham Hoyland found. In extreme conditions, sweating is the real threat because it leads to chilling and can result in frozen clothing. Ventilation is missing in some modern garments, or is often not understood by the user. The 1924 outer-garments were very easily ventilated by relaxing the belt and undoing buttons. Take a look at the clothing chapter in Mark Twight's book, *Extreme Alpinism* (1999, p. 82) and arguably the 1924 clothing layers are closer to Twight's ideals than the common attitudes to layering practised today.

Some Mallory Mysteries Resolved

Viewed through 21st century eyes, the revelation that the 1920s Everest climbers were lightweight pioneers transforms perceptions. (Dennis Lee)

5. The footwear puzzles include why crampons were not used, whether their boots were good enough for the summit and how they kept their feet warm?

Crampons have existed for at least 2000 years, but were crude and ineffective until 10 point crampons were invented by Oscar Eckenstein in 1910. However, early crampons were not as effective as nailed boots in certain conditions, and needed new methods and styles of placing ones' feet.

Crampons were taken on the 1924 Everest expedition but George Mallory did not wear them for his summit bid, relying on his high altitude nailed boots. Nailed boots were far more versatile than modern rubber-soled boots and were very effective on the ice-covered rock that makes up most of the north-east ridge route. Climbers did not use anything as crude as 'hobnails' so often mentioned by journalists in describing old mountain boots. Different nail types and nailing patterns were developed to suit specific conditions and terrain. Nailed boots could and did lead to frostbite, because nails conducted heat away from the feet. However, so did early crampons, their tight bindings often constraining circulation.

It was customary for alpine climbers to select boots large enough to be used with two pairs of socks. For Himalayan climbing the sizing was increased to allow three pairs of socks. We know already that Mallory's lightweight boot was specifically designed to protect against frostbite and that he was wearing the customary three pairs of socks. Without crampons, and with his light boots, the weight on Mallory's lower legs was almost half that of the modern climber.

6. Were these garments better than today?

This research has proved the efficiency of the garments worn by Mallory in 1924 and perhaps more importantly the order and nature of the layers. It has shown that this clothing combination was adequate for the 1924 Everest climbers. It was not, however, superior to today's clothing. Layers, but perhaps more importantly layering techniques, can hopefully be improved in the future because of the information gleaned from this project.

1922 Everest Expedition. The second climbing party descending from their record climb. George Finch and Geoffrey Bruce, as they head back to camp IV at approximately 23,000 feet. Finch, at rear, is wearing his down jacket. (Royal Geographical Society)

Conclusion

As Sir Chris Bonington commented in May 2006:

> While Mallory and Irvine remain mountaineering's greatest mystery, the great thing about this project is that it has shown that they were well enough clad to have had a chance of reaching the summit. It also knocks on the head the popular myth that they were wandering around Everest dressed in tweeds for the grouse moor. They were obviously very well equipped and, indeed, the boots they wore were actually much lighter than the ones we wore when we climbed Everest in 1975!

We still do not know whether George Mallory and Sandy Irvine reached the summit of Mount Everest in 1924, but we do know that the clothing they wore would not have prevented them from doing so. Indeed, members of the 1999 Anglo-American Research Expedition suspected that these men were stronger, fitter and faster than most modern mountaineers and that they were also much better equipped than popular myth suggests. As a result of analysis of Mallory's clothing fragments brought back from Everest and the painstaking work of the replica team from four universities, some of the Mallory mysteries have been resolved and the opportunity for further field tests remains.

Storm on Everest
(Royal Geographical Society)

APPENDIX 1
Mallory Artifacts recovered in May 1999

Clothing

- 3 pieces of beige cotton underwear, shirt, fabric more tightly woven than trouser piece.
- large piece of vest with waist and left sleeve, woven cuff.
- piece of collar labelled *"Hosiers Greensmith, Downes & Son Edinburgh"* (black stitching on white tag).
- 5 pieces of beige-brown silk shirt, largest piece from chest with left sleeve, cuff still buttoned up. Piece of collar labeled *"W.F. Paine, 72 High Street, Godalming"* *"G. MALLORY"* (red stitching on two white tags).
- 3 pieces of dark-brown woollen pullover, largest piece from chest with left sleeve bleached out along edges, waist and cuff with different knitting than chest.
- 3 pieces of pale olive-green silk shirt, largest piece from chest and left side with sleeve. Piece of collar labelled *"Junior Army & [...] Stores Ltd."* (black stitching on white tag).
- 3 pieces (2 small, 1 large) of blue/white striped flannel shirt, large piece of chest and left side with sleeve, some staining on cuff, probably blood.
- pieces of green windproof canvas jacket/ waistcoat :
  a) whole left side of chest down to waist, three button holes and belt loop, side pocket (lined with white cotton), left sleeve and cuff piece lined with black/white striped cotton, small watch pocket on inside (between lower two buttonholes) containing residues of mint cake or lozenge, several blood stains on chest and cuff.
  b) smaller piece from right side of chest down to waist, belt loop, part of right pocket flap.
  c) piece of collar, reinforced with brown linen, coat hanger and label (dark-brown with white stitching, illegible) pieces (25 cm + 40 cm, 5 cm wide) of canvas belt, joined by rusted leather-covered metal buckle.
- 1 triangular piece of cotton underwear, trousers (left hip), beige-white, 25 x 25 x 28 cm.
- 4 cm wide white cotton waistband, sling for braces on outside, label fragment *"Bud [...] Outfitters"* *"G. LEIGH – MA [...]"* (black stitching on white tag).
- 1 piece of red-brown woollen leggings, left hip, 9 cm wide white/brown: cotton waistband on inside, hand-stitched sling for braces on outside.
- 3 pieces of green windproof canvas breeches (matching with jacket):
  a) triangular front piece with 3 buttons, lower button apparently replaced (hand-stitching), labelled *"Burberrys, London"*, one button on left side of waist above hip pocket.
  b) large piece of left thigh down to knee, 4 cm wide white cotton waistband, reinforced with brown linen, hip pocket (white cotton).
  c) small piece of waist, 15x5 cm, with button and sling for braces.

- leather-covered fur wrist band, presumably from gloves,
- 2 flaps (15x5 cm + 7,5x5 cm), joined by 2 ivory-colored buttons.
- oblong piece (12,5x10 cm) of light-brown wool, two layers, stretchy, presumably from scarf.
- yellow-green fingerless glove (left hand).
- fragment of grey wool knitting (3x2 cm), glove?

Footwear and miscellaneous

- light boot (30 cm long, 15 cm high), brown leather with yellow-brown felt lining, abrasions along right side, 5 cm long tear along left side of big toe, white-green discoloration, pieces of faded brown laces remaining in lower right eyelets, inked number 273266 on right inside, sole with nails, 3 nails missing on tip, 1 on either side and 1 on the heel; top part (leather flaps, tongue, laces) of left boot, faded brown-red colour.
- parts of three socks from left foot:
    a) green-violet wool, calf to heel (43 cm long), strongly faded (only heel retaining original colour).
    b) grey-black wool "Walksocken" (heavy wool), calf to heel (40 cm long).
    c) grey-brown wool with white rhombic stitching at upper calf, calf to heel and part of sole (43 cm long), strongly faded (only heel and sole retaining original red-brown colour).
- 2 pieces (25 x 10 cm and 15 x 10 cm) of beige-white woollen putties, seams stitched with small black squares.
- 7 pieces of 9mm three-strand cotton climbing rope (approx. 7,5 m long), red tracer thread.
- sling of yellow-brown webbing (88 cm long, 2,5 cm wide), hand-stitched, twisted.
- piece of blue/white knitting (85 cm long, 1,5 cm wide), knotted, possible Tibetan origin.

Source : We are grateful to Jochen Hemmleb, expedition historian 1999 Mallory and Irvine Research Expedition, for supplying this list from his archive.

APPENDIX 2
Timeline of a century of layering and breathability innovation

1823 Charles Macintosh patented his rubber-coated fabric.
1855 The waterproof jacket developed by Charles Macintosh reached a highly refined stage. The fabric was rubber-coated and the finished garment weighed $5^{1}/_{2}$ oz and could be 'fitted into a cigar box'.
1875 Nares Expedition to Arctic used windproof canvas smocks, the origin of modern layering systems.
1879 Thomas Burberry registered the trademark for his 'self-ventilating cloth', which was to provide polar explorers and mountaineers with protection for the next 50 years.
1923 Haythornthwaite's of Burnley, Lancashire, produced the first batch of a new windproof but breathable fabric known as Grenfell. Grenfell fabric was made for Sir Wilfred Grenfell (a missionary working in Labrador). By the 1930s Grenfell cloth becomes the leading wind/waterproof cotton fabric, supplying all major expeditions and taking over from Burberry.
1938 The beginnings of Teflon® at Du Pont – subsequently leading to Gore-Tex.
1941 Ventile developed by the British Cotton Industry Research Association (the Shirley Institute) perfecting a fabric using long stapled cotton which kept out water when the yarns swelled. The high level of casualties for airmen on convoy duty ditching into icy waters of the Atlantic prompted this research.
1948 Furleen invented as an artificial fur, but was not commercially successful until used by Helly Hansen.
1961 Helly Hansen introduced pile for work wear. It took off for mountaineering in the mid 1970s.
1965 Helly Hansen introduced a range of polypropylene underwear (LIFA®) designed to move moisture away from the skin.
1972 W.L. Gore began experimenting with what became Gore-Tex.
1979 First batch of Pertex manufactured by Perseverance Mills – patent registered 1980. This became the new windproof successor to all previous cotton fabrics above.
1981 Patagonia developed a next step version of pile which they branded Synchilla which Malden Mills later launched as Polartec fleece.

Further Reading

Clothing and Equipment
Mike Parsons and Mary B. Rose, *Invisible on Everest: Innovation and the Gear Makers* (Old City Publishing, 2003)

The Mystery of George Mallory and Sandy Irvine
Conrad Anker and David Roberts, *The Lost Explorer* (Simon & Schuster, 1999)
David Breashears and Audrey Salkeld, *Last Climb: The Legendary Everest Expeditions of George Mallory* (National Geographic, 1999)
Peter and Leni Gillman, *The Wildest Dream: The Biography of George Mallory* (Hodder Headline, 2000)
Jochen Hemmleb, *Detectives on Everest: The 2001 Mallory and Irvine Expedition* (The Mountaineers, 2002)
Tom Holzel and Audrey Salkeld, *The Mystery of Mallory and Irvine* (Pimlico, 1999)
Julie Summers, *Fearless on Everest: The Quest for Sandy Irvine* (Weidenfeld and Nicholson, 2000)
Walt Unsworth, *Everest: The Mountaineering History* (Bâton Wicks, 2000)

Acknowledgements

The research and replication work of this project was led by the brochure authors, under the auspices of the Mountain Heritage Trust (MHT), the owners of the replicas. Thanks are due to Mountain Heritage Trust trustees for constructive comments on drafts of the brochure.

We want to take this opportunity to thank all those involved in bringing the project to a satisfactory conclusion. This, of course, includes the Mallory family and the Royal Geographical Society, the long-term custodians of the Mallory artefacts, for their permission to carry out the project. We are also grateful to the Mount Everest Foundation for their letter endorsing the project. The project could not have been undertaken without the financial support of the Heritage Lottery Fund. The BBC funded the 1999 Mallory and Irvine Research Expedition, initiated by Graham Hoyland and we would especially like to thank Graham for sharing his family's climbing history and Everest experiences with us. Thanks are also due to Pasold Research Fund for its funding and our colleagues at the Institute for Entrepreneurship and Enterprise Development, Lancaster University Management School for their support of the Clothing for Extremes Conferences. Jochen Hemmleb, historian of the 1999 Mallory and Irvine Expedition, has been especially generous with his time and knowledge. Our partners Marian Parsons and Tony Breakell have lived with this project for the last three years and supported us and helped us throughout. Tony, deserves special thanks for his role in the fitting process, so crucial to ensuring the outer garments fitted over the multiple layers and for his role in the unveiling. We are grateful for permission to reproduce images from the collections of the Royal Geographical Society, The Sandy Irvine Trust, The John Noel Collection, The Fell and Rock Climbing Club Library. All images are individually credited. We are also grateful to Everest expert Audrey Salkeld who helped us so much, guiding us through her photograph collection, reading our draft brochure and believing in the project.

Greatest thanks, however, go to the project team who worked so hard and gave so freely of their time and knowledge. The replica team of Amber Rowe, Dave Brook, Vanessa Anderson, and John Angus was drawn from the Textile Conservation Centre, University of Southampton, the Performance Clothing Research Group, University of Leeds, and the Textile Department at the University of Derby and the Performance Clothing MA course at the University of Derby. Joyce Meader, who was responsible for the knitwear, is a freelance historic knitter. The vest was supplied by John Smedley Ltd of Matlock and fabric for the outerwear was made by Woodrow Universal Mills, and supplied free of charge by Burberry. Puttees were hand woven by Simon Young.

About the authors

Mike Parsons was MD of gear-makers Karrimor International from 1960 to 1997. He is now the innovation director for outdoor company, OMM Ltd. In 2003 he was appointed an Honorary Entrepreneurial Fellow of Lancaster University Management School's Institute for Entrepreneurship and Enterprise Development (IEED). Mike is the co-organiser with Mary Rose, of the annual Clothing for Extremes Conference.

Mary B. Rose is Professor of Entrepreneurship at Lancaster University Management School's Institute of Entrepreneurship and Enterprise Development. In 2005 Mary and Mike Parsons won a Lancaster University undergraduate teaching prize for their ground-breaking innovation course. Their book *Invisible on Everest: Innovation and the Gear Makers* (2003) was winner of the 2005 Design History Society Scholarship award and runner up for the Wadsworth Prize in business history in 2004.